NAKAJIMA
Ki-43 HAYABUSA I-III

In Japanese Army Air Force • RTAF • CAF • IPSF Service

Richard M. Bueschel

Schiffer Military/Aviation History
Atglen, PA

Cover artwork by Steve Ferguson, Colorado Springs, CO.

THE RED FALCON'S LAST FLIGHT

In the Papuan campaign of early 1943, the most prominent JAAF unit was the famous 11th Sentai, readily identified by the lightning bolt emblazoned on the tails of their Ki-43 Ia Hayabusa's (Peregrine Falcon), known to the Allies as the Oscar. First Chutai leader Capt. Miyabayashi, the "Red Falcon," is seen here with his wingman in an attack on American C-47s carrying supplies to the forward bases of the allies closing in on Lae. Certain that his Oscar can easily outmaneuver the approaching P-39 escorts, the veteran will fall victim to P-40s of the 49th Fighter Group which have not yet come into view.

After the losses of their armies at Guadalcanal and Buna, the Japanese Imperial Command summoned several Army Air Force units from China to bolster the struggling units in the southern area of operations. The first JAAF group was the illustrious 11th Sentai commanded by LtC. Sugiura which advanced first to Rabaul in support of the evacuation of Guadalcanal, and eventually to Lae, New Guinea, in support of the counter-offensive against Gen. Douglas MacArthur.

The first squadron to reach Lae airdrome was flambouyant Captain Miyabayashi's 1st Chutai. Known to his countymen as the "Red Falcon" due to the gaudy markings of his personal aircraft, the fearless squadron leader often flew lone reconnaissance of the front lines and was never hesitant to engage a superior force. However, the Oscar was not the aircraft to sustain such bravado, and within weeks, both Sugiura and Miyabayashi were dead. The shattered sentai withdrew to refit in their homeland for the ultimate battle in the Philippines in late-1944, there to be annihilated for the last time.

Acknowledgements

Thanks are due to the photo sources cited, with special thanks to Richard L. Seely of Aero Literature for permission to reproduce the photographs from the R. L. Seely Collection; James F. Lansdale for proofing and editing of the original edition for historical accuracy, and James I. Long for bringing the text up to date with current practice nomenclature and model identification.

This title was originally published by Osprey Publications Ltd. in 1970.

Book Design by Mary Jane M_J Hannigan

Published by Schiffer Publishing Ltd.
77 Lower Valley Road
Atglen, PA 19310
Please write for a free catalog.
This book may be purchased from the publisher.
Please include $2.95 postage.
Try your bookstore first.

We are interested in hearing from authors with book ideas on related topics.

Reportedly experimental aircraft Ki. 43-Kai serial number 4313, future production Ki. 43-I aircraft were virtually identical. Shorzoe Abe.

Nakajima Ki-43 HAYABUSA I-III
In Japanese Army Air Force • RTAF • CAF • IPSF Service

In a matter of minutes World War II would come to an end. It was approaching midnight, the 14th of August, 1945, and a P-61 Black Widow of the USAAF 548th Night Fighter Squadron based in the Ryukyus was tracking a night flying Japanese fighter. Lt. Clyde, the pilot, and radar operator Dale Leford visually identified it as an "Oscar" as they forced it closer and closer to the water. They watched it explode as it smacked into a wave. By the time their P-61 touched down again at Ia Shima it was thirty minutes after midnight on the morning of August 15th. The Pacific War was officially over and their gunless "kill" was the last air-to-air score of the war.

It was fitting that the final Japanese Army Air Force loss should be a Nakajima Ki. 43 Type I Fighter Hayabusa (Peregrine Falcon), the "Oscar" of Pacific theater fame. It confirmed the Hayabusa's unique position as one of the very few fighter air-craft of World War II to remain in active combat and production from the first day of the war to the last. It wasn't the Hayabusa's outstanding perfor-mance that kept it in first line service that long. While this may have been the case at the beginning of its combat career in December 1941, the practicalities of desperate need and production ease kept it there over the years. Even the end of World War II failed to terminate the "Falcon's" career, and some were still in service across the expanse of the Orient until the 1950s. This is the story of that long career.

Origins and Initial Production

The outstanding success of the Nakajima Ki. 27 Type 97 Fighter had established chief engineer Hideo Itokawa as the dean of Japanese Army fighter designers. With this one design the Nakajima Air-craft Company (Nakajima Hikoki K.K.) ended

Ki. 43 experimental aircraft serial number 4312 after belly landing during testing. Initial production Ki. 43 followed this basic form. Sekai-no-Kokuki.

Kawasaki's historic role as the prime supplier of Army fighters, a position held by the smaller firm by virtue of consistently winning a series of design competitions. With selection of the Nakajima Ki. 27 design, the JAAF totally rejected the designs of the other two bidding firms for the Type 97 Fighter contract. The timing of this victory was particularly appropriate for Nakajima, for a new attitude toward Army aircraft design competitions was sweeping over Army Air Headquarters in Tokyo. The fantastic cost faced by the manufacturing firms in an open competition where there could be only one winner imposed financial burdens on these producers which could no longer be tolerated by the firms. Political pressure to eliminate the competitive selection of aircraft was applied by Mitsubishi and Kawasaki and reached into the halls of the Japanese Diet, and members of this legislative body passed the word in no uncertain terms to Army Air. Thus, at the very time when the JAAF was considering a follow-on design to their new Type 97 Fighter they were restricted to the selection of a single supplier. At that moment Nakajima stood alone as the producer of modern monoplane fighters for the JAAF. In December of 1937 the firm received a development contract with the secret Kitai (Airframe Number) Ki. 43 designation for a next generation fighter to replace the Ki. 27. By the beginning of 1938 designer Itokawa had assembled a staff of engineers and draftsmen in the Army Design Section at Nakajima and had started preliminary drawings. While the Ki. 43 would ultimately become the lowest numerically designated aircraft in the 40-series of Ki. numbers to reach production, it would be among the last to receive this acceptance. An agonizing three years of alternating failures and successes confronted its designers and production engineers before the full promise of its design was realized.

The initial specs called for an aircraft that was as maneuverable as the Ki. 27, yet capable of greater speed, faster climb and longer range. While adherence to the requirements for the fighter would have given the JAAF the utopian aircraft they were seeking, this performance was beyond the capabilities of the current state-of-the-art. The Ki. 43 was designed by a committee, and it looked it. When the first Ki. 43 prototype, serial number 4301, was completed on December 12th, 1938, the designing

Believed to be an early production Ki. 43-Ia model under service test, this Type 1 Fighter is natural dural with a degraded blue black antiglare panel and white combat stripe. M. Toda.

staff were not particularly proud of their efforts. The design concessions forced upon Itokawa soon backfired, for when Nakajima's own tests were conducted on the prototype at Ojima Airfield at Ota in January 1939 it was obvious the original specs would not be met. Two more experimental aircraft were completed; serial number 4302 in February and 4303 in March and entered the Army test program. It was standard JAAF procedure to order a number of experimental aircraft for evaluation, test them, and then determine if they should drop the project or pursue it.

With the arrival of crack Army pilots used to flying the maneuverable Ki. 27, the Ki. 43 was under trial. The reports were turned in quickly. The Ki. 43 prototypes were stiff, unresponsive to controls, and slow. The landing gear was difficult to retract, the aircraft was too heavy and the cockpit cover was awkward. All of this was particularly hurtful as a modified Ki. 27-Kai had been constructed with a retractable landing gear that bettered the Ki. 43's performance. Following long-winded discussions and reports about the superior maneuverability of the Ki. 27, which most Army pilots were then flying without their cockpit covers, the Ki. 43 was flatly rejected for Army consideration and the matter was considered as dropped.

It doesn't take a manufacturing company long to face a major financial crisis when they are asked to give away a lot of free engineering time and advice. It is the most expensive product they have to sell. With rejection of the Ki. 43 Nakajima was facing trouble, not the least of which was a loss of val fighters. The story of the Ki. 43 would normally have stopped here, except for the persistence of the design and sales staffs of Nakajima. They had too much in the project to give up, and did not intend to abdicate their position as the leading JAAF fighter producer. This time they didn't listen to the Army and set their own goals, and undertook a complete redesign program. The first of ten additional experimental aircraft was completed in November 1939 as serial number 4304, to be followed by nine more, with serial number 4313 completed in September 1940. This time around every trick in the designer's book was applied, with variations in

Initial production Model 1A or Model 1B under service test in the fall of 1941. Aircraft is natural dural overall with two "experimental aircraft" stripes in red. Sora.

enthusiasm in the design staffs. This group had been unfairly hampered by unrealistic demands, and their egos were bruised. One positive thing did come out of the Ki. 43 experience, and that was the apparent success of its new power plant. This was a Nakajima 14-cylinder twin-row radial engine based on current American power plants. It was designated the Ha. 25, and in the same year as the initial tests it was accepted for production as the Nakajima Type 99, 950 h.p. engine, thus beating the production acceptance of its platform aircraft by two years. The Ha. 25 went on to create the famed Sakae series of power plants used on the Mitsubishi A6M Zero-Sen series of Japanese na-

construction, fittings, power plants and armament being tried. The fifth to thirteenth experimental models even mounted advanced Ha. 105 engines instead of the Ha. 25 power plant. A variety of armaments were also tried, with successful variations showing up on production models of the fighter. By the time the twelfth aircraft was being tested in autumn 1940 the design was fairly well frozen and production approval was under consideration. The last of the additional ten experimental aircraft, serial number 4313, cinched the design, and utilized "Butterfly" combat flaps, a technical feature that made the Ki. 43 one of the most dangerous fighters in the air at the start of the Pacific War. First

The same photograph (as seen on page 5) was used by the JAAF for the preparation of Aikoku (Patriotism) cards passed out at local functions to thank the Japanese populace for contributing funds for the purchase of Army aircraft. This example was retouched for Aikoku No. 2203 circa winter 1942-1943. Shorzoe Abe.

mounted on the eleventh aircraft, serial number 4311, this device was uniquely Japanese and was ultimately copied by other nations. These wide chord flaps could be dropped in combat, resulting in a greatly increased controllable wing area. The Ki. 43's maneuverability could thus be instantly enhanced, giving the aircraft a phenomenally tight turning radius and the nimbleness to beat a slower reacting opponent. The JAAF's utopian fighter was now a reality, and on January 9, 1941, Nakajima was officially informed of production approval of the Ki. 43 as the Type 1 Fighter.

Production started at Nakajima's Ota plant in April 1941 with the Model Ki.43-Ia, officially known as the Type 1 Fighter Model 1A, armed with two 7.7mm machine guns. The production run was short lived, with Model 1A serving as a proto-production model to set up the tooling and iron the bugs out of the production line. About 35 were completed by the end of June, followed in July by the Ki.43-Ib, Model 1B, with one 7.7mm gun and a larger bore 12.7mm gun. About 45 of these were produced when the assembly line was converted once again in September, this time to produce the Ki.43-Ic, Model 1C, with two 12.7 mm. machine guns. With that the definitive first production model was off and running. With less than a hundred preliminary mod-

els behind it, the more heavily armed Model 1C was in full production for service use, with a little over 600 being produced by February 1943. Most of the earlier Ki.43-Ia and Ki.43-Ib models were shipped off to the Akeno Army Flying School and other test and training locations while Model 1C production went straight to the combat Air Regiments. The name Hayabusa was given to the combat model to look good in press reports, and by December 8th, 1941, the 59th and 64th Air Fighter Regiments had already received forty of the fighters between them. These service numbers rose rapidly as Nakajima production got into full swing.

The Falcon Enters the Fight

One of the biggest heros to come out of the Sino-Japanese "Incident" was Captain Tateo Kato, commander of the 64th Air Regiment, and a veteran of three years of aerial combat over China, Manchukuo and Outer Mongolia. Kato was also a flamboyant personality who had the loyalty and affection of his men. His young good looks, intelligent eyes and handlebar mustache made good press, and his courage backed up the image. It was only natural that this Super-Hero should have one of the first cracks at the Army's new fighter, so Kato's

Early model Ki. 43-Ia or Ib at Akeno Army Flying School late in 1941. Type 97 Ki. 27 "Nate" fighters used as trainers landing in background. Sora.

64th was slated for initial deliveries. While the press reported Kato's assignment to the Military College in Japan, the 64th was called back to Japan and undertook an extensive program of pre-flight and final flight training on the Ki. 43. There were signs of war in the air, negotiations with the United States were all but terminated, and the 64th slowly became proficient in the use of their new mounts.

Joining the 64th in its aerial adventure was the 59th Air Regiment, known throughout Japan by the flashy lightning stripe markings on its Ki. 27 aircraft. In October of 1941 the 59th was already in training with the Hayabusa at Tachikawa. The 59th had received its first Type 1 Fighters in the late summer and fall of the year. Training was accelerated as the year came to a close and the almost physical and environmental tension of the past months of political activity drew to its ultimate conclusion. You could feel it; there was war in the air. Common Japanese Air Force talk of imminent fighting with the A-B-D Powers (American, British and Dutch) late in November and in the first few days of December 1941 may not have been official, yet it was accurate. Then suddenly, on December 8th, 1941 (Japanese time), in a series of lightning attacks across the broadest fronts in any war in history, Japan attacked at Pearl Harbor, the Philippines, Hong Kong, Thailand and Malaya. Movement of JAAF units into Thailand, including the 59th and 64th, jeopardized the British plan for the defense of Singapore. The march was on, and the Royal Air Force in Malaya and Burma, joined by the American A.V.G., found themselves in Buffalos, Hurricanes and P-40's facing the JAAF Ki. 27 fighters they expected to see, plus a new retractable-gear

fighter they knew nothing about. The result was slaughter. Kato's 64th Regiment, by March 1942 headquartered at Rangoon, became the most famous fighting unit in Asia as their "kill" scores mounted. The Hayabusa was in the fight, and would remain there until this war had come to a close.

As Hayabusa production became fully onstream JAAF units were rapidly re-equipped or newly created. The 1st Fighter Regiment converted from the Ki. 27 to the Ki. 43 by early 1942 in Burma, and later the Dutch East Indies. The 11th, of Nomonhan fame, received their first Ki. 43 aircraft in January 1942, and soon took a heavy load of aerial combat against the Dutch East Indies, American and Commonwealth air forces in the Japanese surge across the Indies. The 24th and 33rd Fighter, and 13th Fighter-Attack Regiments, were converted to the Ki. 43-Ic and were moved into the Indies in the spring of 1942, remaining in New Guinea and the "North of Australia" area until 1944. The 50th and the 77th, both crack Ki. 27 units when the Pacific War began, showed up in Burma and the eastern fringes of India with Hayabusa fighters in February and June of 1942. From this point forward the earlier Ki. 27 Type 97 Fighter became a rarity in the combat zones, with the Hayabusa gaining fame as a tough opponent, and an aircraft with which to avoid a dogfight at all costs.

New units were also formed. The 65th Light Bomber Regiment converted to Hayabusas and settled in for home defence in Chosen, known to the Allies as Korea, and Manchukuo, a Japanese satellite nation made up of the three northern prov-

Wide track landing gear gave the Ki. 43-I the same ground stability as the earlier Ki. 27. Koku-Asahi.

The Type 1 Fighter makes its appearance. This is one of the first photos of the Ki. 43 released to the Japanese and Asian press in January 1942 and shows an early model assigned to the Army Flying School at Akeno. Koku-Asahi.

inces of China and known to the Allies as Manchuria. The Ki. 27 counter-Soviet patrols in Manchukuo were strengthened by Hayabusas of the new 203rd Regiment. The 47th Independent Fighter Company in China received Ki. 43-Ic models prior to its conversion to a regiment in May 1942. The 24th and 71st Independent Fighter Companies, both dispatched to Sumatra in the late spring and summer of 1942, received Ki. 43-Ic fighters as original equipment. The Japanese Army's Falcon had staked out its territorial imperative and looked as if it just might keep all of the other birds away.

It wasn't until the advent of two completely disconnected events within months of each other that the Hayabusa's bubble of invincibility was finally burst. But when the image loss came, it was rapid. Allied pilots soon learned that fast passes at a Hayabusa, with a complete avoidance of close-quarters combat, gave their heavier aircraft an advantage. When a virtually intact Ki. 43-Ic was downed at Chittagong near the Indian border in December 1942, with only the undersides damaged, the Allies finally had a research prize that revealed other weaknesses of the Hayabusa in combat. The captured aircraft, a Hayabusa of the 2nd Company, 50th Fighter Regiment, also revealed much about Japanese nomenclature and regimental markings. Soon many Hayabusa wrecks were being exhibited in India, and flyable examples were under test by the Australians, the Nationalist Chinese and the American Air Forces in China.

The second event was a far more telling one, yet its significance was not recognized by the Allies for some time. The Japanese populace, and specifically the JAAF, felt it immediately. On May 22, 1942, Japan's greatest flying hero, Lieut. Colonel Tateo Kato, having just downed a British Blenheim bomber as his 58th "kill", was suddenly missing in action. He did not return after a sortie over the Bay of Bengal. While it was thought that a British Hurricane shot him down, his fate remains unknown. Thus Japan's national hero disappeared in a shroud of mystery, without physical remains to present to his ancestors. In a catharsis of agony

Early all dural Ki. 43-I at Akeno Army Flying School in the summer of 1942. Inscription forward of stabilizer indicates this is aircraft No. 1029 purchased for the JAAF with funds collected in public drives. Individual number 89 appears in characters above Akeno emblem on rudder. Koku-Asahi.

and repentance the entire Japanese Empire observed his death. The mood of national depression seemed to indicate that the end of one phase of the war was over, and another was about to begin. Kato was elevated two ranks posthumously to an Air Force Major General, his regiment having scored 268 victories under his command. Later in the year a statue of Kato would be erected in the Hall of the War God at the Akeno Army Flying School as an inspiration to future JAAF pilots. The timing of Kato's death almost seemed to be based on clairvoyance, for one month later the Battle of Midway was over and the course of the war had changed.

Trio taking off at Akeno. Koku-Asahi.

the airframe in order to improve its performance and capability. The first of what eventually led to five Ki. 43-II test models was completed at Nakajima's experimental shops at Ota in February 1942 just as the Ki. 43-Ic was proving itself superior to the archaic pre-war RAF, Royal Australian, American, Dutch and other Allied aircraft in Burma and the Dutch East Indies. Many small details were refined or added on the new Ki. 43-II, including an optical gun sight, modest pilot armour of 13 mm. steel, an improved radiator, re-engineered wing tips, and a complete re-routing of the air intake systems to accommodate the powerful new 1120 h.p. Nakajima Ha. 115 power plant selected for the second-generation Hayabusa, the latter an advanced

Trio of early Type 1 Fighters over Akeno Army Flying School, summer 1942. Camouflage on the Ki. 43 in the foreground indicates prior use by an operational unit. Koku-Asahi.

The Hayabusa Comes of Age

The theory behind obsolescence-at-delivery is that any production item can be improved, with the newer model stepping in to maintain production as the earlier model has reached its peak of usefulness. This is particularly true of well engineered aircraft, and can be continued again and again as long as the airframe shows promise. The Ki. 43 was the JAAF's best example of such renewable design vitality. As the Model 1C Hayabusas flowed into combat, experiments were being conducted on

In the air. Koku-Asahi.

Landing. Koku-Asahi.

development of the Ha. 25 series. The room was also needed for the addition of a two-stage supercharger. As the five prototypes were modified, re-evaluated and tested between February and May 1942, plans were made to convert the Nakajima production lines at Ota to the first Ki. 43-IIa production model without undue delay. Certain specifics were frozen into the design, including the armament of two Ho. 103 Type 1 machine guns of 12.7 mm., plus provisions for two 30 kg. bombs and up to two 250 kg. bombs for special needs. The airframe revisions and wing strengthening required to carry this arms load remained with the Ki. 43 design throughout the remainder of its career, making this the standard armament for all future production models. As the years progressed this armament became increasingly inadequate, yet it remained until the very last Hayabusa model. In a masterpiece of production engineering the Ki. 43-IIa was phased into the lines at Ota with the first production issue in November 1942. Within sixty

In press release Japanese censors removed the Akeno Army Flying School emblems from photos of the Type 1 combat trainers. Compare to same photograph with the rudder markings in place. Sekai no Tsubasa.

days the Ki. 43-IIa rate of production was running smoothly and exceeding that experienced with the Ki. 43-Ic. While most were standard models, a substantial number were also built as tropical models with the addition of a large filtering system, evidence of Japan's new commitment in South-East Asia. Nakajima continued to produce the Ki. 43-IIa model until September 1944 when it dropped Hayabusa production to concentrate on the newer Ki. 84.

Recognition of the Hayabusa's role as the JAAF's standard fighter for the foreseeable future also dictated the need for alternate sources of supply, with both Tachikawa and the 1st Army Air Arsenal also located at Tachikawa (and commonly referred to as Rikugun) in line for production orders. The Army Air Arsenal was also slated for Ki. 43-IIa production, with Tachikawa held in reserve for a later model. The Rikugun facility theoretically produced its first Ki. 43-IIa in October 1942, finishing

Akeno Army Flying School Ki. 43-Ic photo retouched to remove school emblems. Type 1 Fighter in rear carries combat stripe, not noted by retoucher, while the school emblem was removed. Richard L. Seely.

seven by the end of the year, but the facts of the matter were far less encouraging. Nothing seemed to work right. The jigs didn't match, and the workers were inexperienced. By the end of the year only four patched up examples were actually ready for delivery, and things just got worse the next year. The whole idea was dropped in November 1943 with 49 of the poorly-assembled fighters having been completed.

Distribution of the Ki. 43-IIa and its tropical version to JAAF units began company by company in the spring of 1943, with units up to strength by the coming of summer. The "North of Australia" 1st, 13th, 24th, 26th and 33rd Regiments all received the Ki. 43-IIa, but now things had changed. Japan's

The new Ki. 43-I "trainer" at Akeno. Ki. 27 Type 97 "Nate" trainer variant without spats in background, biplane Ki. 9-Kai Type 95 Trainer "Spruce" in foreground. Koku-Asahi.

Another angle from the same photo shoot with markings of the Type 1 at Akeno remaining. Rear Ki. 43 is camouflaged with combat stripe. Koku-Asahi.

island positions were on the defensive, and new Allied fighters now outclassed the Hayabusa in speed and armament. The Ki 43-IIa models also showed up with the 25th Air Regiment in occupied China, a number of the Home Island Defense units in Japan, including the 54th and 63rd Air Regiments, and in the famous 64th Air Regiment in Thailand, now renamed the "Kato Regiment." The combat experiences of late 1942 made it quite evident that the changes being made in the Hayabusa were not keeping up with the times. In June 1942, even as the Ki. 43-IIa model was finalized for production, Nakajima experimented with a further modification of the Model 2. By the end of August, and three more experimental aircraft, the basic concepts of the Ki. 43-IIb model were established. Slated for Tachikawa production the Ki. 43-IIb had its wings clipped and a simple jet effect exhaust mounted in place of the open exhaust of the Ki. 43-IIa model. These design tricks boosted the speed by almost 40 km/hr., making the Model 2B a better performer than its Model 2A sister of a few months earlier. Tachikawa became the sole producer and began production in May 1943. The early fall of 1943 saw the Ki. 43-IIb Hayabusa start to enter service in virtually every combat zone patrolled by the JAAF.

Fighting for Its Life

The days of the Richthofen-like regimental commander floating protectively overhead, watching his Japanese "Army Eagles" destroy hapless enemy opponents, were over in the Pacific. Replacing them for a now harried JAAF were slashing attacks against jungle bases in New Guinea in which bombs were falling and strafing attacks were all but over before the Japanese pilots could even get to their aircraft. In Thailand and Indo-China and north into China itself, Allied bombers were miraculously bombing Japanese installations while leaving local facilities intact. Over Burma the battle was a brutal give-and-take. North to Manchukuo the potential battle with the Soviet Union was always there, but the game was a waiting one. And over Japan the JAAF held its breath. Nothing yet. But somehow the inexorable return to Japan and the need to defend the homeland became more of a reality day by day.

Japanese defense was just beginning to firm up late in 1943 and early in 1944, a time-consuming task considering the Doolittle Raid took place almost two years earlier. New units equipped with the Hayabusa, such as the 20th Air Regiment at Lake Otami on Hokkaido; the 51st and 248th at Ozuki; the 52nd at Bofu and the 71st at Kameyama, Shimane, plus re-equipment of other units with the Hayabusa such as the 54th Air Regiment on Hokkaido and the 47th Air Regiment moved in from China to protect Tokyo, made the new Ki. 43-IIa and Ki. 43-IIb fighters important in Japan's defense. Across the Sea of Japan a large force-in-being was tied up in Manchukuo awaiting any possible Russian moves. Hayabusas became the prime fighter of the Counter-Soviet Patrol, creating an unused inventory of well-equipped fighter forces that the JAAF would soon draw on to patch holes in the enormous defensive quilt of Japan's newly won empire. New Hayabusa regiments were formed and stationed in Manchukuo, including the 30th and the 26th, the latter at Eimonton, and both formerly flying light bombers. New fighter units created from scratch included the 48th and 203rd Air Regiments. Older Ki. 27 units still serving in Manchukuo, including the 59th, 68th and 77th Air Regiments, were also re-equipped.

In Korea the 65th Air Regiment received Hayabusas, as did the 25th and 31st Air Regiments in China opposing the American 14th Air Force there. In Burma and Thailand the newer Hayabusa models reached the 21st, 50th and 64th Air Regiments. But it was in the South-West Pacific area that the greatest changes were made, for it was here that the combat lines were being clearly drawn. As Allied pressures increased additional shifts were made to check the Japanese rate of retreat. In the beginning the battle tested 24th and 33rd and 63rd Air Regiments in New Guinea, plus the 11th and 13th Hayabusa Air Regiments at Rabaul, seemed capable of holding the line with the normal rate of attrition and the support of other fighter units in the

1930s technology in a '40s fighter. The rigors of combat didn't allow much time to peer through the scope to sight the enemy. The Ki. 43-IIa replaced this troublesome tube with an optical gun sight. Richard L. Seely.

area. But the dam broke as General MacArthur's "Island Hopping" tactics began to take their toll. Suddenly the JAAF's positions across New Guinea and the "North of Australia" area were virtually defenseless and fighter cover was desperately needed. The JAAF scoured its resources and frantically shifted units into New Guinea that were ill-equipped for jungle fighting. Japan's defense perimeter was weakened in a series of emergency

moves that in themselves failed. While the JAAF had a difficult enough time getting the transferred units into New Guinea to save the situation, when things got worse they couldn't get them out. Units pulled out of Manchukuo included the 26th and the veteran 77th, the latter moved first to Burma and then immediately to New Guinea. The 68th was also moved after conversion from Hayabusas to the newer Kawasaki Ki. 61 fighters. Japan itself pro-

Retractable gear Ki. 43-Ic at Akeno, summer 1942. Older stiff legged Ki. 27 Type 97 Fighters also used as combat trainers in background. Richard L. Seely.

vided the 248th Hayabusa Air Regiment. Yet by the end of 1942 it was hard to find a Japanese fighter over New Guinea, and by the spring of 1944 the once powerful JAAF was a litter of weed-filled wreckage, its aircraft smashed on the the ground before they got a chance to take off. The destruction of the Hayabusa units was so total that complete Air Regiments disappeared from everywhere but the JAAF records in Tokyo. The 77th never flew as a unit again, its wreckage still to be found in the jungles. The 248th and 68th Air Regiments were officially disbanded at JAAF Headquarters in Tokyo on August 20, 1944. Other regiments barely got officers and men out of the trap, leaving their Ki. 43 aircraft behind them. The JAAF attempted to redress part of this balance by more aggressive tactics in the Burma area, attacking the "Hump" route of American transports bringing supplies into China from India. Early in 1944, Japanese bombers escorted by Hayabusas attacked the Allied air bases in Assam. Lone Hayabusas intercepted slow C-46 and C-47 transports in the rarefied air over the mountains. JAAF inventories in Burma were supported by stealing needed aircraft from Manchukuo, Chosen and China, keeping these cannibalized units at low strength. Then, on March 7th, 1944, an entire JAAF force of twenty fighters and eighteen bombers was wiped out by American P-40 and P-51 fighters, with the exception of one returning Hayabusa. The shattered pilot told the story of the battle and his survival over Radio Tokyo in what must be considered as a major propaganda goof. By May 1944, as the Japanese drove hard toward

A friendly tussle at the Akeno Army Flying School. Type 1 Fighter, Model 1, left, tangles with a Type 97 Fighter in a mock dogfight. These tactics served the JAAF well in the early months of the Pacific War, but were useless when command of the air was lost.

Kohima, sweeps of up to thirty Hayabusas over the Imphal area supported the Japanese ground forces. It was Japan's last significant offensive, for the front ultimately collapsed and the JAAF was again on the run.

And then things went from bad to worse. By now American bombers were over Japan. The loss of closer islands would mean that the bombers would soon be accompanied by fighters. And the newer, faster and now maneuverable Allied fighters were knocking down the nimble Hayabusa in greater numbers. Yet the Ki. 43 series still remained numerically the most important Army fighter in Japanese service due to its uncomplicated long production runs and model improvements. The invasion of the Philippines, and the defense of Japan, would both find the Hayabusa in use in greater numbers than any other single Army fighter type, primarily because there were so many. These final tests of Japan's resolve would ultimately sap the JAAF of its strength, while individual Hayabusa units and pilots would demonstrate great courage and skill against a far stronger enemy.

The Last Hayabusas

Nakajima's production interest in the Ki. 43 ended with the conversion of the Ota lines to the later Ki. 84 Hayate Type 4 Fighter "Frank." The Hayabusa increasingly became a Tachikawa project. Late models of the Ki. 43-IIb, beginning around July 1944, had single exhaust stacks that further improved performance. These Ki. 43-II-Kai models led the way to the last production model, the Ki. 43-IIIa mounting the more powerful Nakajima Ha-115-II radial of 1190 h.p. After testing of some Nakajima experimental aircraft, the Ki. 43-IIIa entered production at Tachikawa in October 1944 with only a slight drop in the Hayabusa production rate that was picked up within thirty days. Over a thousand of these models eventually reached service. Tachikawa also experimented on its own with a further model designed for Home Island Defense that would carry two 20-mm. cannon and have a more powerful Mitsubishi Ha.112-II engine of 1290 h.p. Only two of these clip-winged Ki. 43-IIIb experimental models were built, and testing was just under-

Arming a Type 1 at Akeno. Hiko-Nippon.

way when the war ended. The Ki. 43-IIIa remained the definitive production model as Japan defended the remnants of its imperial holdings.

The JAAF's master plan to block American invasion and eventual control of the Philippines Islands began to take physical shape in the late summer of 1944. By the time the American landing had taken place at Leyte on October 19th, shifts had been made throughout the JAAF to strengthen the

Philippines. Regiments were pulled out of Guinea, Burma, Thailand and China and sent to the 4th Air Army in the Philippines with their Hayabusas often supported by some of the new Ki. 84 Hayate fighters, the Hayabusa replacement. These units included the 1st, 11th, 13th, 24th, 31st, 33rd and 65th Air Regiments, as well as the 24th Independent Fighter Company then in Sumatra. Hayabusa regiments transferred out of Japan included the 20th from Formosa, the 51st, 52nd, 54th, 63rd, 73rd and 246th out of the home islands, as well as the hastily-created 71st and 72nd Air Regiments. Most of these units reached the Philippines in late October or mid-November 1944 to face an enemy stronger in numbers and equipped with superior aircraft. The Japanese losses were staggering, and when some of the surviving units were rotated back to Japan the following spring only the pilots returned, their aircraft having been abandoned or lost in the battle. Some of the units never made it, and were terminated in the records months after the campaign. The 11th, 72nd and 73rd Air Regiments, among others, were completely decimated and useless as viable units. Typical was the 30th Air Regiment, transferred twice in October and November 1944, virtually destroyed in December, and finally returned to Japan as a skeleton force in May 1945.

Elsewhere new challenges were facing the JAAF, with the 23rd Hayabusa Air Regiment putting up a short but strong defense at Iwo Jima in February 1945, followed by the toughest fighting of

Training conversion flight at Akeno, July 1942. Koku-Shonen.

the Pacific War at Okinawa beginning in April. Here the 13th Air Regiment, pulled out of Celebes; and the 52nd, 65th, 101st, 102nd and 204th Hayabusa Air Regiments, among others, fought hard and well against Allied opposition. Lieutenant Ryotaro Kamibo, top scoring Japanese "ace" of the Okinawa campaign, with over seventy reported "kills", was a Hayabusa pilot. Many unnamed and unknown Japanese "aces" earned their titles at Okinawa, only to be killed in action and to remain anonymous. The Hayabusa also made its debut as a Taiatari type at Okinawa, becoming one of the most used Army aircraft for suicide missions.

The final bastion was Japan itself. The Hayabusa had proved to be notably unsuccessful against the American Boeing B-29 Superfortress, yet many remained on station as day and night fight-

The working end of the Ha. 25 powered Ki. 43-Ic at Akeno, summer 1942.

Beauty of line. A conversion training Model 1B climbs over Akeno. Koku-Asahi.

the 71st in the Western Defense Sector; with others including the 30th, 51st, 52nd, 54th, 65th, 101st, 102nd and 103rd somehow surviving the Philippines and Okinawa campaigns to fight again in the skies over Japan. In Manchukuo the 48th, 104th, 203rd and 204th Air Regiments still flew Hayabusas to face the Soviets and Outer Mongolians in their more modern Lavochkin and Yakovlev fighters in a battle that lasted less than a week before Japan finally capitulated and ended the fighting on August 15th, 1945.

Foreign and Post-War Use

No Pacific War aircraft was more widely used or more widely distributed than the Nakajima Hayabusa series. The result of this geographic distribution was that the aircraft survived the war in some numbers and continued to be used after the defeat of Japan.

Prime foreign user of the Hayabusa, and the only air force to which the aircraft was originally exported, was the Royal Thai Air Force. As a member of the so-called Greater East-Asia Prosperity Sphere and a co-belligerent of Japan, the maintenance of a separate air force by Thailand was of great propaganda value while a somewhat limited tactical asset to Japan. In the spring of 1944 a Company of former JAAF Ki. 43-IIb Hayabusa fighters

ers for use against escorting fighters and the Allied carrier aircraft then beginning to appear in numbers over Japan. Hayabusas at the various training schools throughout Japan also doubled as defense fighters when an alert sounded in their areas. It was in these battlefields that the Hayabusa saw its last major combat. Home Defense units flying the Hayabusa included the 20th, 112th and 246th Air Regiments in the Middle Defense Sector; the 23rd and 47th in the Eastern Defense Sector;

Japanese censor has removed tail markings for this Ki. 43-Ic photo prior to press release. Base is Akeno. Hiko-Nippon.

Model 1C Hayabusa coming in for a landing at Akeno. The Ki. 43-Ic was to the JAAF what the Bf. 109 was to the Luftwaffe, the Hurricane to the Royal Air Force and the P-40 to the U. S. Army Air Corps.

was transferred to the Royal Thai Air Force to replace its obsolete American Curtiss Hawk-III and II-75N fighters. This also gave the Japanese the opportunity to infiltrate the RTAF, already recognized as a hotbed of pro-Allied feeling within the Thai armed forces, and a group that had little sympathy with Japan's cause. JAAF instructors worked with the American-educated Thais, but they were soon dismissed to leave the Thai pilots on their own. Ultimately, the RTAF flew a squadron of the Hayabusa fighters in southern China under JAAF command, while others were retained for the defence of Bangkok where they rarely took to the air.

When the war ended the Thais rushed their aircraft through a normalization process, removing all evidence of wartime paint and insignia. The wartime Thai insignia made use of the Japanese Hinomaru on the upper wing surfaces, something

the Thais would prefer not to remember in the postwar years, and the traditional Thai white elephant on a red panel on the undersides of the wings and the sides of the vertical tail. The historic pre-war Thai roundel insignia was applied to the surviving Hayabusa fighters by the first week in September, 1945, the aircraft themselves now a natural dural. These fighters remained in first-line service until the late 1940's when they were replaced with more modern British and American types.

In the Dutch East Indies the Hayabusa became the first fighter of the newly-formed Indonesian People's Security Force, a nationalistic outgrowth of an underground group that gained strength during the Japanese occupation. Opposed to renewed Dutch control, the insurgent force built up an air force out of wrecks found in the former JAAF dump at Djakarta and elsewhere in the islands. A num-

A later combat camouflaged Model 1C at Akeno with the school markings eliminated by the Japanese censor. Hiko-Nippon.

Introducing the combat Hayabusa to the Japanese and Asian public. Highly retouched photo released in January 1942 shows a lineup of the aircraft of the 64th Air Regiment. Hiko-Nippon.

All markings were removed from the Ki. 43-Ic at the time of the January 1942 Hayabusa press introduction. Richard L. Seely.

ber of Hayabusas, probably belonging to the JAAF 33rd Regiment stationed at Medan in the Indies when the war ended, served with the Indonesian forces between 1946 and 1949.

A number of Hayabusa fighters were flying in Nationalist Chinese colors after the war had ended. Others were captured by the People's Liberation Army of China forces as they spread into Manchuria in October 1945. They were used sporadically by the PLAAF in China's Civil War.

Possibly the most unconventional user of the Hayabusa was the Republic of France. Faced by a rising Viet insurgency as they returned to French Indo-China, and having few aircraft on hand to quell

Sand-and-spinach camouflage of the early service models assigned to the 64th Air Regiment can just be made out. Hiko-Nippon.

the disturbances, the French utilized confiscated Japanese aircraft in the area. The last Hayabusas of the once-proud "Kato Regiment" and others were soon impressed into French service, being flown by Groupes de Chasse I/7 and II/7 in counter-in-surgency operations. They were quickly replaced by more modern American aircraft flying under French colors.

Published as a photograph of Lt. Colonel (posthumously Major General) Tateo Kato's actual 64th Air Regiment mount at the time of his death in May 1942, this is actually a Ki. 43-Ic Hayabusa of the 50th Air Regiment. Hiko-Nippon.

The Nakajima Ha. 25 was the most powerful engine yet used on an Army fighter. Hiko-Nippon.

The Imperial Japanese Navy introduced their new Zero-Sen fighter in December 1941. The Army scrambled and countered this interservice press rivalry in January 1942 with a Hayabusa introduction event. Photographs are highly retouched to remove 64th Air Regiment unit markings. Hiko-Nippon.

Refueling the Ki. 43-Ic at its press introduction. Hiko-Nippon.

Right: The JAAF started the Hayabusa with the same old fashioned gasoline powered truck mounted hub spinners they had used on their Type 95 Kawasaki biplane and Type 97 Nakajima fixed gear monoplane fighters in China since the middle 30's. Richard L. Seely.

One of the most famous Hayabusa photos, widely circulated and reprinted in the Greater East Asia Co-Prosperity Sphere news media. These are Model 1C Hayabusas of the 1st Air Regiment over French Indo-China early in 1942. Hiko-Shonen.

The perfect platform for a "dogfighting" air force, the Ki. 43-Ic was arguably the most maneuverable monoplane fighter of World War II. 50th Air Regiment. Richard M. Bueschel.

The 11th Air Regiment flies its Ki. 43-Ic Hayabusas over the Dutch East Indies in summer 1942. This action photograph was wire transmitted to neutral nations and widely reproduced throughout the world during the Pacific War. Richard M. Bueschel.

The Ki. 43-Ic at war. Hayabusa of the commander, 2nd Company, 64th Air Regiment, at a north Malayan airfield as the unit worked its way south. Koku-Asahi.

Another Ki. 43-Ic of the 64th Air Regiment in Malaya, spring 1942. Late model captured car under the wing. Koku-Asahi.

Malay peninsula, March 1942. A Hayabusa of the 2nd Company, 64th Regiment. Koku-Asahi.

The Model 1C Hayabusa fans out into the field. Japan's rapid military advancement south brought the cold weather JAAF into tropical climates. Crews dressed for the occasion. Richard L. Seely.

Hayabusas of the 50th Air Regiment, 2nd Company, over Burma-India area, spring 1942. Early in the Pacific War this unit was highly publicized and its lightning stripe markings were often illustrated in the Japanese press. It was some time before the Allies recognized the importance of these unit markings. Richard M. Bueschel.

The 50th Combat Regiment warms up the Ki. 43-Ic's of the 2nd Company in Burma late in 1942. Note the absence of an outline around the fuselage Hinomaru insignia. Hiko-Shonen.

Ki. 43-Ic in Burma in 1943. Unit is the 50th Air Regiment. Color of company marking appears on cowling and spinner. Richard M. Bueschel.

Two Hayabusas of the 50th Air Regiment in close formation. As they did in the Ki. 27 "Nate," JAAF fighter pilots often flew the Hayabusa with the canopy slid back and open.

The 50th Air Regiment over Burma, April 1942. Koku-Shonen.

Pilots of the 50th Air Regiment return from a mission and discuss their encounters with "A, B" (American and British) aircraft back on their airfield in Burma early in 1942. Regimental Ki. 43-Ic Hayabusas in background. Richard L. Seely.

Model 1C Hayabusa banking in flight. Tip of fuselage marking by wing root suggests 50th Air Regiment in Burma. Richard L. Seely.

The original of the photo reportedly showing Colonel Kato's fighter aircraft of the 64th Air Regiment was actually taken from this shot of two 50th Air Regiment Hayabusas over Burma in spring 1942. Hiko-Nippon.

Right: Deadly view. Hayabusa closing in for a "kill." Richard L. Seely.

Getting started at the Army Aviation Maintenance School at Tokorozawa, spring 1943. The maintenance school had examples of all active JAAF equipment for training purposes. Hideya Ando.

Hayabusas in combat formation. The performance of these aircraft made remarkable aerobatic teams possible. JAAF "Red Falcons" gave public displays at air shows in Japan. Richard L. Seely.

Ex-operational sand-and-spinach camouflaged Ki. 43-Ic over Akeno in January 1943. Individual aircraft marking 17 is in white at the top of the rudder. Full combat markings, including white stripe, are applied. Koku-Asahi.

By the late summer of 1943 most Hayabusas at the Akeno Army Flying School were splotch camouflaged to double as defense fighters in the event their use was necessary, a precaution slowly initiated after the Doolittle Tokyo Raid of April 1942. These are Ki. 43-Ic models. Koku-Asahi.

Field maintenance in Burma, November 1942, 64th Air Regiment. Asahigraph.

Captured Ki. 43-Ic test flown by the Nationalist Chinese in the spring of 1943. P-5017 markings applied to the fin indicate the Hayabusa is a "Pursuit" aircraft of the Chinese Air Force (CAF). Jack Canary.

Another view of the newly delivered Ki. 43-IIa model to service units, June 1943. Koku-Shonen.

Ki. 43-IIa in the war zone. Hayabusa Model 2A carries the markings of 1st Company, 64th Air Regiment, Thailand, summer 1943. Richard L. Seely.

The Model 2A enters service, spring 1943. Unit unidentified. Hiko-Nippon.

Fussa Airfield, Yokota, hosts a variety of aircraft. Ki. 43-Ic in foreground, Kawasaki Type LO "Thelma" transport version of the American Lockheed 14 appears through the landing gear legs, and a Nakajima Type 100 "Helen" heavy bomber is under the wing. Richard M. Bueschel.

Arming the beast. The Ki. 43-IIa Model 2A improved performance and made the pilot's task easier with an optical sight. Armament was increased with heavier bore guns. Richard L. Seely.

Ki. 43-IIa fighters in service, June 1943. Koku-Shonen.

Aikoku (Patriotism) Aircraft No. 2068 purchased with public funds collected in wartime subscription drive. Aircraft is Ki. 43-IIb. Hideya Ando.

Ki. 43-IIa of the 3rd Company, 64th Air Regiment, taxies into its base in Thailand, summer 1943. M. Toda.

Ki. 43-IIa Model 2A combat trainers at Akeno were camouflaged to double as defense fighters if needed. Hiko-Nippon.

Left: "Bandage" markings on the aft fuselage later used for home defense units were first tried out at Akeno late in 1942. Koku-Asahi.

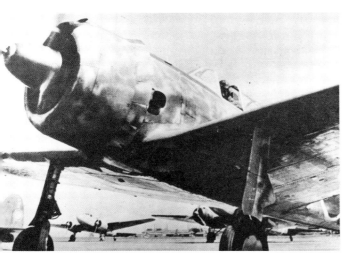

Line-up at Akeno, full combat markings. M. Toda.

Left: Akeno received Ki. 43-IIa aircraft for conversion as soon as unit deliveries began. Tachikawa Type 1 Advanced Trainer "Hickory" navigational trainer can be seen in the background. Seiso Tachibana.

Ki. 43-IIa at the Tokorozawa Maintenance School. Aircraft was flown, disassembled, re-assembled and maintained, and flown again in a repeating cycle of instruction for ground crews. Hideya Ando.

Headquarter's company Hayabusa of the 25th Air Regiment ready for take-off in China. The 25th saw a great deal of action against Chennault's 14th Air Force in SW China. Robert C. Mikesh.

The combat camouflaged Ki. 43-IIa fighters at the Akeno Army Flying School in late 1943 no longer had the early war combat stripes. Hideya Ando.

Company Commander's Ki. 43-IIa Hayabusa of 2nd Company, 25th Air Regiment in China, 1944. Markings are red with white edging. Hideya Ando.

Chinese occupation force flew Ki. 43-IIa. This is aircraft 55, 1st Company, 25th Air Regiment in summer 1943. Splotched camouflage over dural. M. Toda.

Aircraft 20, 2nd Company, 25th Air Regiment in China. White spinner. Richard L. Seely.

Life in the combat zones with the Ki. 43-IIa Model 2A, July 1943. Hiko-Nippon.

Quiet days in Burma. Ki. 43-IIa of 1st Company, 50th Air Regiment, center, being readied for a patrol mission. 2nd Company aircraft in foreground and rear. Hiko-Nippon.

Still from the 1943 wartime aviation movie "Hayabusa Regiment." Aircraft painted like those of the 64th "Kato Regiment" in Thailand were used for realism. Seiso Tachibana.

Another still from the wartime movie "Hayabusa Regiment" shows cockpit details, utilizing a static Model 1 aircraft set up in a studio. Seiso Tachibana.

Typical movie heroics from the 1943 wartime film "Hayabusa Regiment" now has the pilot in a Ki. 43-IIa Model 2A. This is a film worth finding and enhancing with a translated sound track for VCR, laser disc or CD-ROM viewing. Seiso Tachibana.

Taking off with the new Ki. 43-IIa Model 2A equipment of the 2nd Company, 25th Air Regiment, summer 1943. Unit markings were later changed to a slanted line. Koku-Asahi.

Conversion training of 25th Air Regiment, early markings, summer 1943. Koku-Asahi.

Night fighter variants of the Ki. 43-IIa were pressed into service over Japan in the fall of 1944 after the beginning of night raids by American B-29's. Fuel tanks extended time over target cities. Koku-Shonen.

Hayabusa of the 33rd Air Regiment, New Guinea, at the end of 1943. This unit was moved to the Philippines the following September. White symbol on vertical tail indicates aircraft is of the 1st Company.

By summer 1944 about all that remained of the 248th Air Regiment was wreckage strewn across the jungles of New Guinea. This Ki. 43-IIb, flown by the Regimental Commander, was found at Alexishafen Airfield. Vertical tail marking is seven white or yellow "double ones" (2+4+8) across the fin and rudder. Robert C. Mikesh.

The way they were. Ki. 43-IIa-Kai tropical captured in New Guinea. Drop tanks outward of the landing gear extended range, provided the aircraft was able to get off the ground. This one didn't. Warren E. Woolmann.

Burned out hulk of a Ki. 43-IIb Hayabusa of the 33rd Air Regiment on the airfield at Hollandia, New Guinea, in May 1944. 2nd Company marking is red over jungle green camouflage. James F. Lansdale.

Wreckage at Hollandia, New Guinea, shows the tail sections of two Model 2B Hayabusas of the 248th Air Regiment. The "double ones" unit markings are in yellow over green camouflage. James F. Lansdale.

Wreckage of tropical Ki. 43-IIb of the 63rd Air Regiment, a unit completely obliterated in New Guinea. Tail unit of 78th Air Regiment Ki. 61 Type 3 Fighter "Tony" in background. Photo taken April 22, 1944 at Korako airstrip, south of Aitape. USAAF Official.

Tucked under the palms at Hollandia, New Guinea, May 1944, strafed and scrapped Model 2B Hayabusa of the 33rd Air Regiment sits uselessly. 2nd Company red marking on vertical tail is outlined in white. D. G. Cooper via James F. Lansdale.

Above: Tropical Ki. 43-IIa-Kai Hayabusa believed to be of 3rd Company, 63rd Air Regiment, in New Guinea. Jack Canary/James I. Long.

Below: Full rear view of the crashed Ki. 43-IIa-Kai in New Guinea shows skinny fuselage and large horizontal tail surfaces. Jack Canary/James I. Long.

Above: New Guinea wreckage shows weathering of splotched camouflage over dural to good advantage, closely approximating an all-dural appearance. Jack Canary/James I. Long.

Below: The 63rd Air Regiment was originally posted to Home Island Defense, and was then transferred to New Guinea in 1943 where it saw much action and was wiped out. Jack Canary/James I. Long.

To the victor goes the goodies. Hayabusa Model 2A captured in New Guinea, 1944. White markings suggest 1st Company of the 33rd Air Regiment, a highly decorated unit so decimated in New Guinea it was stricken from JAAF records and officially disbanded in May 1945. Richard L. Seely.

New simplified unit markings of the 77th Air Regiment showed up in New Guinea late in 1943. 1st Company white flash markings across fin and rudder carried the individual aircraft identification on the upper rudder, here stripped off as a souvenir at Hollandia, May 1944. D. G. Cooper via James F. Lansdale.

2nd Company, 59th Air regiment Hayabusa on the airfield at Hollandia, New Guinea, May 1944. American souvenir hunters have stripped off the rudder markings, leaving only the white outlined red stripe on the fin. D. G. Cooper via James F. Lansdale.

This classic photograph of the Japanese Army Air Force Nakajima Ki. 43-IIb Model 2 Hayabusa Type 1 Fighter is not what it appears to be. It is an aircraft rebuilt by Technical Intelligence SWPA and flown in Japanese markings for a series of films and photographs that were used in British and American wartime IFF training and recognition manuals. Richard L. Seely.

Ki. 43-IIb with drop tanks, unit unknown. Sekai-no-Kokuki.

Strafed Ki. 43-IIa Model 2A of the 25th Air Regiment, assigned to the Chinese occupation force. Unit was formed in China late 1942 with the Hayabusa as original equipment. Richard L. Seely.

Japan had their own "Rosie The Riveters" at work on the aircraft production lines as shown on the cover of the November 10, 1943 issue of the wartime aviation magazine Shashin Shuho. Richard L. Seely.

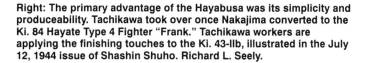

Right: The primary advantage of the Hayabusa was its simplicity and produceability. Tachikawa took over once Nakajima converted to the Ki. 84 Hayate Type 4 Fighter "Frank." Tachikawa workers are applying the finishing touches to the Ki. 43-IIb, illustrated in the July 12, 1944 issue of Shashin Shuho. Richard L. Seely.

The Hayabusa was a diminutive aircraft by world standards, and it took some doing for the home front production workers to get into the tight spots of the after fuselage. Working women were a major social and cultural change for wartime Imperial Japan. Richard L. Seely.

Hayabusas of the 1st Company, 25th Air Regiment in central China in the summer of 1944. Hiko-Shonen.

The three photos above show Hayabusa production lines at Tachikawa were well established, and running smoothly, until the bombings began. Fuselage sections were held up on saw horses, wings mounted in an assembly room, with final assembly on moving jigs with landing gear extended. Richard L. Seely.

Hayabusas assemble for unit transfer to the Philippines, fall 1944. Shashin-Shuho.

Right: The Instructor's Company at Kumagaya Army Flying School was equipped with Ki. 43-IIb. Fuselage "bandage" was yellow. Robert C. Mikesh.

Left: Kumagaya conversion trainers in full Home Defense markings doubled as Hayabusa interceptors. Robert C. Mikesh.

Combat-weary Hayabusas of the 1st and Headquarter's Companies, 64th Air Regiment, in Thailand late in 1944. Robert C. Mikesh.

Hayabusa Ki. 43-IIb models could carry long range fuel tanks or up to two 250kg. bombs for long range or fighter-bomber missions. Splotched camouflage is well shown here. Time is July 1944. Hiko-Nippon.

Model 2B Hayabusa of 2nd Squad, Instructor's Company, Kumagaya Army Flying School, early 1944. Yellow "trainer" insignia "bandage" panel over natural dural on after fuselage. Ki. 84 Type 4 Fighter "Frank" fighters in background. Robert C. Mikesh.

Combat ready. Japanese Home Island Defense, spring 1945. N. Saito.

Slim, trim, beautiful in the air, but sadly obsolete by fall 1944. A Ki. 43-II-Kai Hayabusa of 1st Company, 64th Air Regiment, flies over Thailand. Seiso Tachibana.

Fighter-bomber Ki. 43-IIb aircraft of the 64th Air Regiment over Thailand, February 1944. Koku-Shonen.

Ki. 43-II-Kai equipped as a fighter-bomber in the Philippines, late 1944 or early 1945, unit unknown. Seiso Tachibana.

Hayabusa units often flew aircraft with a variety of camouflage schemes. Highly retouched photo shows 50th Air Regiment operating in the Imphal area in June 1944. Shashin-Shuho.

Ki. 43-II-Kai of 1st Company, 1st Air Regiment at Clark Field in the Philippines as found by American forces in late 1944. Unit was attached to the 12th Air Brigade of the 4th Air Army. Robert C. Mikesh.

Wing mounts could be used for drop tanks or bombs. Unit unknown. James Tindal.

The Hayabusa was the most numerically important Japanese Army fighter in the Philippines, and was slaughtered there. Seiso Tachibana.

A Ki. 43-Ic warms up. This photograph was taken in Japan in 1944 at a Home Defense base and demonstrates the front line use longevity of the early Hayabusa models. Hideya Ando.

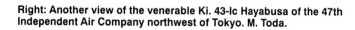

Right: Another view of the venerable Ki. 43-Ic Hayabusa of the 47th Independent Air Company northwest of Tokyo. M. Toda.

Combat ready Hayabusa of the renamed Akeno Air Training Division takes off for a daylight interception. Aircraft in the foreground are Akeno based Ki. 44 Type 2 "Tojo" fighters. Shashin-Shuhu.

Above: The 47th Independent Air Company (later redesignated the 47th Air Regiment) flew Ki. 44 Type 2 Shoki "Tojo" fighters, rear, in Tokyo defense in autumn 1944, as well as its Ki. 43-Ic fighters, foreground. M. Toda.

Below: A combat veteran of many years, this Hayabusa Model 1C of the 47th Independent Air Company is shown in full air defense markings at Narimasu Airfield, northwest of Tokyo. M. Toda.

Home Defense Ki. 43-IIIa Model 3A Hayabusa carries white "bandage" markings on wings and fuselage. M. Toda.

"Headless" Hayabusa stacked by American occupation forces at Kiro, Japan, in May 1946 prior to burning. Markings are of Akeno Air Training Division. Holmes G. Anderson.

Model Ki. 43-II-Kai of the 19th Army Special Attack Squadron, Middle Defense Sector, Itami, Hyogo, Japan, in winter 1944-45. M. Toda.

The last flight! Late model Hayabusa with a 500kg. bomb takes off on a Taiatari (suicide) mission. Within the hour the pilot in this picture had died. Shorzoe Abe.

Former JAAF Ki. 43-IIb fighters were turned over to the Royal Thai Air Force in late spring 1944. Former Hinomaru circular area under the wing has been replaced by the rectangular Thai white elephant insignia on a red panel. Thai fighter unit flew beside JAAF units. Koku-Asahi.

Wartime Thai insignia was ancient Royal white elephant on red field which appeared as a red "spot" for identification, making it similar to the Japanese insignia. For absolute positive identification Thai upper wing surfaces carried the normal Japanese Hinomaru marking in red or dark orange, the latter fading to pink. Koku-Asahi.

Hayabusa Ki. 43-II-Kai in the air. Jet effect exhaust stacks gave an extra kick to the top speed and paved the way for the Model 3. Richard L. Seely.

Line-up of the 48th Air Regiment at Kiangwan Airfield, Shanghai, China, at the end of the war. Photo was taken by U. S. Army photographer on September 8, 1945. Hayabusa fighters are lined up in the center. Aircraft at far right is a Manshu Ki. 79 Type 2 Advanced Trainer, a two-seat conversion of the Ki. 27 "Nate" produced in Manchukuo. U. S. Army Official.

Above: The Hayabusa became an experimental test bed for a wide variety of speed and performance improving innovations. Here an air chewing four bladed propeller is tried on a ski equipped Model 2A to search for improvements in cold weather operations. Richard L. Seely.

Above right: Some tests led to field modifications to meet special needs. Model II-Kai in Manchukuo is equipped with retractable skis. Aircraft is probably with the 48th or 203rd Air Regiment of the Counter-Soviet Patrol which saw brief action in August 1945. Richard L. Seely.

Above: The last complete Hayabusa! The Ki.43-IIIb was engineered and built by Tachikawa. Only two were completed and testing was underway when the Pacific War ended. Scheduled for production in 1946. Sekai no Kokuki.

Left: The end of the line. Production Ki. 43-IIIa fuselages found at the Tachikawa plant at the end of the war, quickly crushed and junked. Holmes G. Anderson.

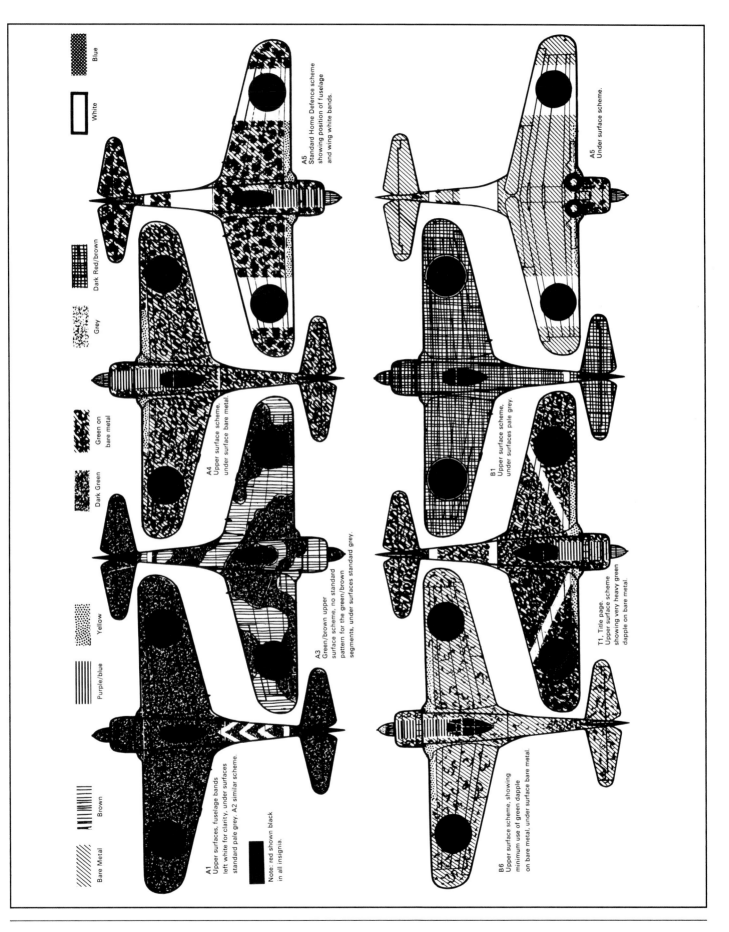

Blue

White

Dark Red/brown

Grey

Green on bare metal

Dark Green

Yellow

Purple/blue

Bare Metal

Brown

A1
Upper surfaces, fuselage bands left white for clarity, under surfaces standard pale grey. A2 similar scheme.

Note: red shown black in all insignia.

A3
Green/brown upper surface scheme, no standard pattern for the green/brown segments, under surfaces standard grey.

A4
Upper surface scheme, under surface bare metal.

A5
Standard Home Defence scheme showing position of fuselage and wing white bands.

A5
Under surface scheme.

B1
Upper surface scheme, under surfaces pale grey.

T1, Title page.
Upper surface scheme showing very heavy green dapple on bare metal.

B6
Upper surface scheme, showing minimum use of green dapple on bare metal, under surface bare metal.

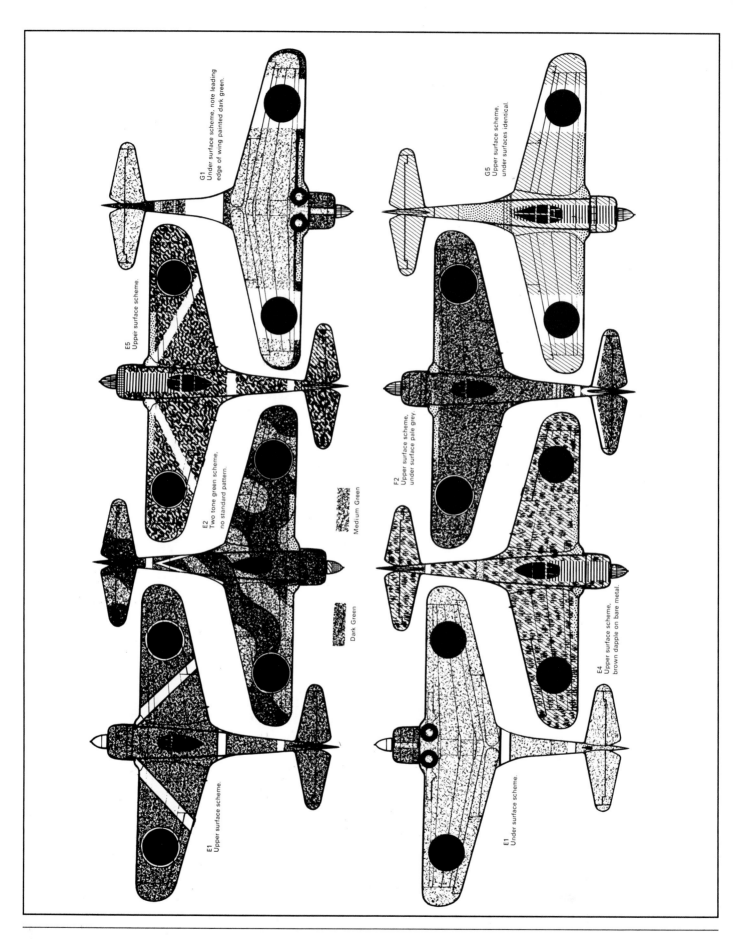

G1
Under surface scheme, note leading edge of wing painted dark green.

E5
Upper surface scheme.

E2
Two tone green scheme, no standard pattern.

E1
Upper surface scheme.

Medium Green

Dark Green

G5
Upper surface scheme, under surfaces identical.

F2
Upper surface scheme, under surface pale grey.

E4
Upper surface scheme, brown dapple on bare metal.

E1
Under surface scheme.

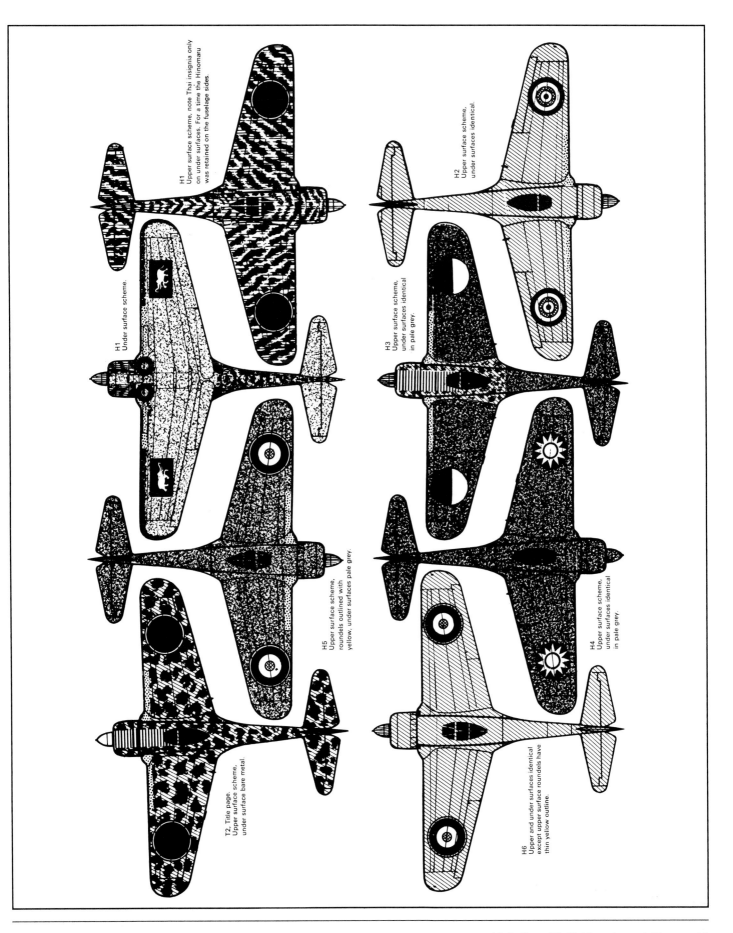

H1
Upper surface scheme, note Thai insignia only on under surfaces. For a time the Hinomaru was retained on the fuselage sides.

H1
Under surface scheme.

H2
Upper surface scheme, under surfaces identical.

H3
Upper surface scheme, under surfaces identical in pale grey.

H5
Upper surface scheme, roundels outlined with yellow, under surfaces pale grey.

H4
Upper surface scheme, under surfaces identical in pale grey.

T2, Title page.
Upper surface scheme, under surface bare metal.

H6
Upper and under surfaces identical except upper surface roundels have thin yellow outline.

Nakajima Ki-43 Hayabusa I-III   49

AIR REGIMENTS

Regiment	When used	Area of Operations	Former A/C	Later A/C	Comments
1st Fighter	Early 1942- early 1945	French Indo-China, Burma, Dutch East Indies, North of Australia-Area, Rabaul, Philippines	Ki.27	Ki.84	Saw much combat. Remnants pulled out of the Philippines. Unit officially disbanded in July 1945.
2nd Reconnaissance	1942-July 1945	China, SW Pacific	Ki.27	Ki.84	Unit officially disbanded in July 1945.
11th Fighter	Jan.1942- Nov. 1944	Malaya, French Indo-China, Dutch East Indies, Burma, Rabaul, Philippines	Ki.27	Ki.84	Saw much combat. Virtually annihilated in the Philippines.
13thFighter-Attack	April1943-end of war	Japan (Home Defense) North of Australia, Area, New Guinea, Rabaul, Dutch East Indies, Celebes, French Indo-China, Philippines, Okinawa, Formosa	Ki.10 Ki.27	Ki.84 Ki.45	Saw much combat. Unit frequently decimated and rebuilt.
20th Fighter	Dec.1,1943- end of war	Japan (Hokkaido) Philippines, Formosa, Okinawa, Japan (Middle Defense Sector)	None	Ki.84	Unit formed at Itami, Hyogo on Dec.1, 1943 with Ki.43 as original equipment. Assigned to night fighter Operations in late 1944.
21st Fighter	Oct.15,1942- Dec.1944	French Indo-China Dutch East Indies, Malaya, Burma	Ki.27	Ki.84 Ki.45	84th Independent Fighter Company and 82nd Independent Light Bomber Company reformed as a fighter regiment at Hanoi, French Indo-China, on Oct. 15, 1942.

50

Regiment	When used	Area of Operations	Former A/C	Later A/C	Comments
23rd Fighter	Oct.11,1944- end of war	Iwo Jima Japan (Eastern Defense Sector)	None	Ki.44 Ki.61	Former Training Company reformed at Inba Chiba, Japan, on Oct. 11, 1944 with Ki.43 as original equipment.
24th Fighter	March 1942- late 1944	Dutch East Indies, North of Australia Area, New Guinea Philippines, China	Ki.27	Ki.45	Saw much combat.
25th Fighter	Oct. 20,1942- end of war	China, Korea	None	Ki.84	Chinese occupation force. Unit formed at Kanko, China, on Oct. 20, 1942 with Ki.43 as original equipment.
26th Fighter	Oct. 2,1942- end of war	Manchoukuo, China Philippines, French- Indo-China, Sumatra, North of Australia Area New Guinea, Formosa	Ki.51b	Ki.61	Formerly Light Bomber Regiment. Unit reformed at Eimonten, Manchoukuo on Oct. 2, 1942 with Ki.43 as original equipment. Saw much combat.
30th Fighter	February 1944- May 30, 1945	Manchoukuo, Phillppines, Japan (Home Defense)	Ki. 30 Ki. 51b	None	Formerly Light Bomber Regiment. Unit reformed in Manchoukuo in February 1944 with Ki.43 as original equipment, Officially disbanded May 30, 1945.
31st Fighter	February 1944- May 30,1945	China French Indo- China, Philippines, Thailand	Ki 30 Ki.51b	None	Formerly Light Bomber Regiment. Unit reformed in China in February 1944 with Ki.43 as original equipment. Unit disbanded May 30, 1945.
33rd Fighter	Spring 1942- May 1945	French Indo-China, Dutch East Indies, North of Australia Area, New Guinea, Phillppines, Sumatra	Ki.10 Ki.27	Ki.61	Saw much combat. Decimated in the Dutch East Indies and officially disbanded in the records in May 1945.

Regiment	When used	Area of Operations	Former A/C	Later A/C	Comments
47th Fighter	May 1942-late 1944	Japan (Eastern Defense Sector)	None	Ki.44 Ki.84	Formerly 47th Independent Fighter Company. Reformed as 47th Regiment In May 1942. Retained some Ki.43 aircraft.
48th Fighter	Nov. 5, 1943-end of war	Manchoukuo	None	None	Unit formed at Botankoshu, Manchoukuo, on November 5, 1943 with Ki.43 as original equipment. Counter-Soviet patrol. Unit disbanded at Taiken, Manchoukuo, at war's end.
50th Fighter	Feb.1942-early 1945	Burma, Thailand, French Indo-China, Philippines, Formosa	Ki.27	Ki.84	One of first units to receive Ki.43. French Indo-China and Thailand Occupation Force.
51st Fighter	April 28,1944-end of war	Japan, Philippines Japan (Home Defense)	None	Ki.84	Unit formed at Ozuki, Yamaguchi, Japan, on April 28, 1944 with Ki.43 as original equipment.
52nd Fighter	April 28,1944-end of war	Japan, Philippines Okinawa, Japan (Home Defense)	None	Ki.84	Unit formed at Bofu, Hofu, Yamaguchi, Japan, on April 28,1944 with Ki.43 as original equipment.
54th Fighter	Late 1942-end of war	Japan (Hokkaido Defense), Eastern China Kurile Island, Philippines, Karafuto, Sakhalin	Ki.27	None	Flew Ki.27 until late 1942 when replaced by Ki.43. Unit had four years of home island defense. Disbanded at Sappore, Ishikari, at war's end.
59th Fighter	Autumn 1941-Summer 1943	Japan, French Indo-China, Malaya, Dutch East Indies, Manchoukuo	Ki.27	Ki.61 Ki.100	Saw much combat. One of two units to enter the Pacific War with Ki.43 equipment on service test.

52

Regiments	When used	Area of Operations	Former A/C	Later A/C	Comments
63rd Fighter	Feb. 25, 1943- end of war	Japan (Hokkaido) New Guinea, Philippines	Ki.27	None	Saw much combat. Unit formed at Hachinohe, Amori, on Feb. 25, 1943 with Ki.27 and Ki.43 as original equipment.
64th Fighter	Summer 1941 - end of war	Japan, French Indo-China, Malaya, Sumatra, Java, Burma, Thailand	Ki.10 Ki.27	Ki.44 Ki.84	Saw much combat. Reportedly first unit to receive the Ki.43 and one of two units to enter the Pacific War with Ki.43 equipment. Used Ki.43 longer than any other JAAF unit. Regimental Commander Lt./Col. Tateo Kato was killed over the Bay of Bengal and became a national hero.
65th Fighter	July 1941- early 1945	China, Philippines, Formosa, Okinawa, Japan (Home Defense)	Ki.32 Ki.51b	Ki.45 Ki.61	Formerly Light Bomber Regiment. Unit reformed in China in July 1941 with Ki.43 as original equipment.
68th Fighter	May 1942- March 1943	Manchoukuo	Ki.27	Ki.61	Replaced Ki.27 with Ki.43 in Manchoukuo. Later converted to Ki.61 in March 1943 and sent to New Guinea where it was all but annihilated.
71st Fighter	June 30,1944- end of war	Philippines, Japan (Western Defense Sector)	Ki.51b	Ki.84	Formerly Light Bomber Regiment. Hastily re-formed as Fighter for Philippines defense on June 30, 1944 with Ki.43 and Ki.84 as original equipment.
72nd Fighter	May 1944- May 1945.	Japan, Philippines	None	Ki.84	Unit hastily formed for Philippines defense. Official formation was June 30, 1944. Sent to Philippines in December. Virtually annihilated there. Unit officially disbanded May 30, 1945.
73rd Fighter	Sept.17.1943- May 1945	Japan, Philippines	None	Ki.84	Unit formed at Kashiwabara at Gifu, Japan, on Sept. 17, 1943. Assigned to Philippines defense in December 1944 and virtually annihilated there. Officially disbanded May 30,1945 for the record.

Regiments	When used	Area of Operations	Former A/C	Later A/C	Comments
77th Fighter	June 1942-Aug. 20,1944	Manchoukuo, Burma, Dutch East Indies, New Guinea	Ki.10 Ki.27	None	One of best known JAAF Regiments. Saw much combat. Completely annihilated in New Guinea and officially disbanded for the records at Hollandia, New Guinea, on August 20,1944.
101st Fighter	Sept. 1944-end of war	Okinawa, Japan (Home Island Defense)	None	Ki.84	Unit officially formed at Kameyama, Shimane. Japan, on Nov. 10, 1944 with Ki.43 and Ki.84 as original equipment.
102nd Fighter	Sept.1944-July 30,1945	Okinawa, Japan (Home Island Defense)	None	Ki.84	Formed at Kameyama, Shimane, Japan, along with 101st Fighter on Nov. 10. 1944. Ki.43 and Ki.84 both original equipment. Losses led to transfer of 102nd remnants to the 103rd Fighter Regiment on July 30, 1945.
103rd Fighter	Sept.1944-end of war	Yura, Awaji Islands	None	Ki.84	Unit formed at Kameyama, Shimane, Japan, on Aug. 25, 1944. Expanded by remnants of 102nd Fighter Regiment in Summer 1945.
104th Fighter	Nov. 30,1944-end of war	Manchoukuo	None	Ki.84	Counter Soviet Patrol. Unit formed at Heizan, Manchoukuo, on Nov. 30, 1944. Fought against Russian and Mongolian aircraft in last week of war.
112th Fighter	July 10,1945-end of war	Japan (Middle Defense Sector)	None	Ki.84	One of last new Regiments. Formed at Komaki, Gifu, Japan, on July 10, 1945 with Ki.43 and Ki.84 aircraft. Unit in process of being moved to Gifu City when war ended.
203rd Fighter	April 1942-end of war	Manchoukuo	None	None	Counter Soviet Patrol. Unit formed in Manchoukuo in April 1942 with Ki.43 as original equipment. Fought less than a week in Soviet and Mongolian invasion in August 1945.

Regiments	When used	Area of Operations	Former A/C	Later A/C	Comments
204th Fighter	July 20, 1944-end of war	Japan (Kyushu), Burma, French Indo-China, Japan (Home Island Defense) Formosa, Manchoukuo	Ki.27	None	Formerly Bomber Regiment flying Ki.27 fighters Unit reformed at Chinzei, Kyushu, Japan, on July 20, 1944 with Ki.27 and Ki.43 as original equipment.
246th Fighter	June 1943-April 1945	Japan (Osaka) Philippines, Japan (Middle Defense Sector)	Ki.27	Ki.84	Home defense unit transferred to Philippines in November 1944.
248th Fighter	Aug. 10,1942-Aug. 20,1944	Japan (Ozuki), New Guinea	None	None	Unit formed at Ozuki, Yamaguchi, Japan, on Aug 10,1942 with Ki.43 as original equipment. Annihilated in New Guinea. Officially disbanded in New Guinea by JAAF headquarters, Tokyo, on August 20,1944 for the record.

55

Independent Companies

Company	When used	Area of Operations	Former A/C	Later A/C	Comments
24th Fighter	May 1942- Jan. 1945	Sumatra, Philippines	None	Ki.84	Saw much combat. Unit transferred to Phillipines in Oct. 1944.
47th Fighter	Sept. 15, 1941 - May 1942	China, Malaya	None	Ki.44	Unit formed at Canton, China on Sept. 15, 1941 with Ki.43 and experimental Ki.44 aircraft as original equipment. Known as the "Kingfisher Company." Later reformed as 47th Fighter Regiment.
71st Fighter	July 1941 - end of war	Japan, Sumatra, Andaman Islands, French Indo-China, Malaya	None	None	Saw much combat. Unit reportedly disbanded in May 1945, although Ki. 43 remained in use with 71st Fighter until end of war.

Flight Training Companies

Company	When used	Location	Former A/C	Later A/C	Comments
39th	Spring 1945- end of war	Yokoshiba Airfield	Ki.79	Ki.61	Used in Home Island Defense in addition to training.

56

Flight Drilling Companies

Company	When used	Location	Former A/C	Later A/C	Comments
1st	July 1944- end of war		None	None	Unit formed in July 1944 when Drilling Companies were established.
2nd	July 1944-end of war		None	None	As above.
4th	July 1944-end of war		None	None	As above.
5th	July 1944-end of war		Ki.27	Ki.61	As above.
13th	Nov.1944-end of war		None	None	
14th	Oct.1944-Dec. 1944		None	None	Unit disbanded in December 1944.
17th	Oct.1944-end of war		None	None	
19th	Oct.1944-end of war		None	None	

Training Schools

School	When used	Location	Former A/C	Later A/C	Comments
Akeno Army Flying School	Nov 1941-June 1944	Akeno, Hitachi	Ki.27	Ki.44	First flying school to initiate Ki 43 pilot training. Reformed in June 1944 to create both the Akeno Air Training Division and Hitachi Air Training Division.
Akeno Air Training Division	June 1944-end of war	Akeno	None	Ki.44	Former Akeno Army Flying School. Unit formed in June 1944 with Ki.43 and Ki.44 equipment.
Kumagaya Army Flying School	1942-1944	Kumagaya	Ki.27	None	Joined Akeno AFS for Ki.43 flight training due to Pacific War JAAF expansion.
Army Aviation Maintenance School	Early 1942-end of war	Tokorozawa	All JAAF aircraft	All JAAF aircraft	Supplied to Army Aviation Maintenance School for Ki.43 maintenance and repair training.

Taiatari (Suicide) Regiments

Regiment	When used	Area of Operations	Former A/C	Later A/C	Comments
1st Special Attack	Summer 1945	Japan (Honshu)	None	Unknown	Regimental headquarters and facilities at Mito Airfield, Honshu, Japan, in last few months of War.
19th Special Attack	Winter 1944-1945	Japan (Hyogo)	None	Unknown	Regimental headquarters and facilities at Itama, Hyogo, Japan.

58

Foreign Service

Country and Unit	When used	Area of Operations	Comments
Royal Thai Air Force (Thailand)	Late 1943-1948	Thailand, South China Malaya, French Indo-China Thailand	Standard Royal Thai Air Force fighter during Pacific War. Remained in service as fighter until 1948.
Republic of China Air Force (Nationalist China)	1943-1948	China	Captured examples test flown by Nationalists during Pacific War. A few captured examples flown by Nationalists during Chinese Civil War, 1946-1949.
Red Army Air Force	Oct.1945-July 1946	Manchuria, North China	Former JAAF aircraft acquired in Manchuria (Manchoukuo) and North China by the Chinese Army (Communist China). Flown by Red Army Air Force until Chinese Communist military forces were reformed in July 1946.
People's Liberation Army Air Force (Communist China)	July 1946-1949	North China, Central China	People's Liberation Army Air Force formed in July 1946. Sporadic use of Ki.43 along with Ki.44 and Ki.84 in Chinese Civil War, 1946-1949.
Air Army of France	Aug 1945-1946	French Indo-China	Former JAAF aircraft confiscated in French Indo-China after Japan's surrender. Flown by Groupes de Chasse I/7 and II/7 against Viet insurgents until arrival of later American aircraft.
Indonesian People's Security Force	Sept.1945-1949	Dutch East Indies, Indonesia	Former JAAF aircraft. Flown by Indonesian Nationalist pilots against Netherlands East Indies Air Force.

NOTE: These lists are not to be regarded as complete as only those for which Ki.43 use has been confirmed have been identified.

SPECIFICATIONS
Nakajima Ki.43 Type 1 Fighter Hayabusa (Peregrine Falcon)

Note: All dimensions in Original Japanese metric. Dimensions and climb in meters (m), weights in kilograms (kg), distances in in kilometers (km) and speeds in kilometers – per hour (km/hr). Data in parenthesis are estimates or approximate.

Model and Specs	Ki.43 Experimental	Ki.43-*Ka* Experimental	Ki.43-Ia	Ki. 43-Ib	Ki. 43-Ic	Ki 43-II Experimental
Span (m)	11.5	11.437	11.437	11.437	11.437	11.437
Length (m)	8.80	8.832	8.832	8.832	8.850	8.920
Height (m)		3.270	3.270	3.270	3.270	3.273
Wing Area (m²)		22.000	22.000	22.000	22.000	22.000
Weight Empty (kg)			1,580	(1,600)	1,924	
Weight Loaded (kg)			2,048	(2,100)	2,100	
Weight Loaded Max. (kg)			2,243		2,580	
Max. Speed (km/hr)			495/ 4000m	495/ 4000m	490/ 4000m	
Crusing Speed (km/hr)			320/ 2500m	320/ 2500 m	320/ 2500 m	
Climb (m/ min.)				5000/ 4' 50"	5000/ 15' 12"	
Armament/M.G. (mm)	2x7.7	2x7.7*	2x7.7	1x7.7 1x12.7	2x12.7	2x12.7
Armament/Cannon (mm)	—	—	—	—	—	—
Armament/Bombs (kg)	—	—			2x15	—
Power Unit Manufacturer	Nakajima	Nakajima	Nakajima	Nakajima	Nakajima	Nakajima
Type	Ha.25	Ha.25**	Ha.25	Ha.25	Ha.25	Ha.115
H.P.	950	950	950	950	950	1120
Aircraft Manufacturer	Nakajima	Nakajima	Nakajima	Nakajima	Nakajima	Nakajima
First Built	Dec.12, 1938	Nov. 1939	April 1941	July 1941	Oct. 1941	Feb. 1942
Number Built	3	10	(35)	(45)	716†	5

* Experimental aircraft serials Ki.4310 and Ki.4313 had 2 x 12.7 mm M.G.
** Experimental aircraft serials Ki.4305 and Ki.4313 had Nakajima Ha.105 of 1100 h.p.
† All models Ki. 43-I series
†† All models Ki.43-IIa series. Rikugun built 49.
††† Included in Ki.43-IIb totals. Ki.43-IIc designation often used is incorrect.

Ki 43-IIa	Ki 43-IIa-*Kai* Tropical	Ki 43-II-*Kai* Experimental	Ki 43-IIb	Ki 43-II-*Kai*	Ki 43-III Experimental	Ki 43-IIIa	Ki 43-IIb
11.437	11.437	10.837	10.837	10.837	10.837	10.837	10.500
8.920	8.920	8.920	8.920	8.920	8.920	8.920	
3.273	3.273	3.085	3.085	3.085	3.273	3.273	
22.000	22.000	21.400	21.400	21.400	21.400	21.400	21.400
1,975		1,729	1,975			2,040	
2,642			2,413			2,725	
			2,642			3,060	
515/ 6000m	(490)		548/ 6000m			555/ 6,100m	
			345/ 4000m				
5000/ 6' 20"						5000/ 5' 19"	
2x12.7	2x12.7	2x12.7	2x12.7	2x12.7	2x12.7	2x12.7	2x12.7
—	—	—	—	—	—	—	2x20
2x30 2x250	2x30	2x250	2x30 2x250	2x30 2x250	2x250	2x30 2x250	2x30 2x250
Nakajima	Nakajima	Nakajima	Nakajima	Nakajima	Nakajima	Nakajima	Mitsubishi
Ha.115	Ha.115	Ha.115	Ha.115	Ha.115	Ha.115-II	Ha.115-II	Ha.112-II
1130	1130	1150	1150	1150	1190	1190	1290
Nakajima Rikugun	Nakajima	Nakajima	Tachikawa	Tachikawa	Nakajima	Tachikawa	Tachikawa
Oct.1942	Spring 1943	June 1942	May 1943	(July 1944)	May 1944	Oct. 1944	June 1945
2524††	(300)	3	1531	(500)†††	(3)	1098	2

Mitsubishi A6M-1/2/-2N ZERO-SEN
in Japanese Naval Air Service
Richard M. Bueschel

Reknowned Japanese aircraft historian Richard Bueschel revises and updates his classic series of books on Japanese Naval and Army Air Force aircraft of World War II. The A6M-1/2/-2N ZERO-SEN is the first volume. All variations and markings are covered in this the first of a projected multi-volume series.

Size: 8 1/2" x 11"   64 pages, over 150 photographs

ISBN: 0-88740-754-4   soft cover $14.95

JAPANESE AIRCRAFT
Code Names & Designations
Robert C. Mikesh

From ABDUL to ZEKE, this handbook covers all Allied designations for Japanese Navy/Army aircraft of the Second World War . Each aircraft is presented alphabetically according to its code name, and is also cross-referenced to its official (long) designations and project (short) designations. Also covered are the non-code named aircraft, and a listing of popular names of Japanese Navy and Army aircraft.
Size: 6" x 9" 192 pages, over 170 b/w photographs
ISBN: 0-88740-447-2   soft cover $14.95

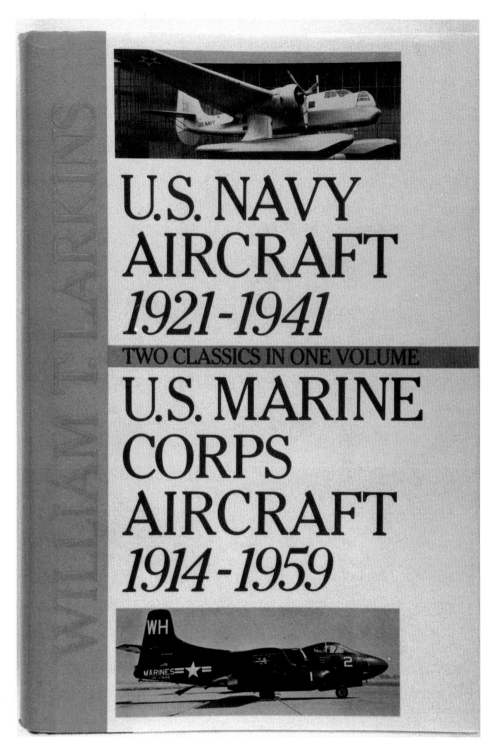

U.S. NAVY AIRCRAFT/U.S. MARINE CORPS AIRCRAFT
Two Classics in One Volume
William T. Larkins

More than thirty years after their first publication, William T. Larkins's illustrated surveys of Navy and Marine Corps aircraft remain two of the most valuable books about these airplanes ever written. First published in 1959 and 1961, these two volumes have become classics eagerly sought in the rare-book market by aviation buffs worldwide. They have become the criteria against which serious aviation research is measured. With over 1,000 photographs combined, they remain the definitive record of the formative years for Navy and Marine Corps aviation. Larkins's emphasis throughout is on squadron use, experimental and one-of-a-kind types, insignia, colors and marking schemes, technical innovations, and the service duty and tactical deployment of the various aircraft.

Size: 6" x 9"   608 pages, over 1000 black and white photographs
ISBN: 0-88740-742-0   hard cover $39.95